SALT

Where it is and What it does

By

Suzanne C. Smith R.N.

NOTICE

This book is intended as a helpful reference book only. Not as a medical guide or manual for self-care or treatment. If you have medical problems, please seek medical attention, do not wait. The information here is written to help you get the best care possible. This book is not a substitute for your personal nursing service or medical provider. This book can, however, prepare you to interact more effectively with healthcare service providers. And be a better shopper and consumer of healthy foods.

Third Edition

Copyright © 2021 All rights reserved.

ISBN-13: 978-1975895525

ISBN-10: 1975895525

Dedicated to all those who have an interest in achieving their best possible health. Press on! Information is powerful. You will see opportunities to make small changes- which may yield significant results!

Contents

1 Salt- It's critical   pg 10

2 Hidden salt   pg 17

3 Prepared foods   pg 24

4 Food fixes   pg 28

5 High blood pressures   pg 33

6 Salty Facts   pg38

References and Resources pg 42

About the Author   pg 48

1 Salt-It's Critical

Your body, and the bodies of all animals, must have sodium chloride-salt-in order to be alive. Our bodies cannot manufacture its own salt.

However vital it is, we still must consider: Some people are sensitive to salt.

Studies show that a sensitivity to salt can almost double the risk of dying from a heart attack. It is estimated one out of four people is salt-sensitive. What that means is: under clinical observation, an increase in the consumption of salt will cause the blood pressure to go up. That is salt sensitivity.

Close to 400,000 deaths in the US each year are attributed to high blood pressure according to the Centers for Disease control and prevention (CDC). If we as a culture decrease our sodium intake thousands of lives annually may be spared. Is salt worth dying for?

You are reading this book because you may suddenly have a keen interest in SALT. If you have read my other books, you have already learned that information is powerful. Now, gain power over salt.

SALT, what we put on the table, has a chemical name-SODIUM CHLORIDE. It is a natural element found all over the world.

SODIUM, one of the chemical elements found in salt, is a mineral. CHLORIDE is a mineral ,both are electrolytes.

This book will help you **get power** over SALT.

- We will see what SALT is
- Where SALT is
- How SALT affects our bodies.

I hope you are motivated to help improve your health and the health of others.

You can be sure this information will be given to you in common English-not "medical-ese". You will not have to get out your high school chemistry book to decipher what you will read.

So, relax. This will be interesting. I promise.

The average, normal human body is approximately 0.4 percent salt. (This is very close to the amount in sea water). That is like four out of a thousand, or four pennies out of ten dollars.

The body has many "salts"- but for the sake of our discussion we will focus on Sodium Chloride-common table salt. The body must contain sodium chloride (chemically written Na+ Cl-). It is necessary for many body functions. Salt is vital to human life. Every cell in the body contains salt. An average adult's entire body contains about 250 grams of salt, which is about 8.8 ounces.

Some things that salt does:

Maintains proper fluid balance in the body

Helps nerves to send and receive information, Sodium is critical for the activity of all nerve signals within our bodies. The nerve messages to and from the brain require salt.

It helps remove acidity from body cells and fluids.

All muscles, including the heart, need it to contract (tighten) and to relax (loosen).

It helps to prevent muscle cramps.

A brief overview.

So, you have an interest in learning more about SALT (Na+Cl-). Perhaps you've just left the doctor's office. Maybe you've been told "eat less salt". But where do you start?

After studying organic chemistry, nutrition and anatomy let me spare you that hard work. There is a mountain of details. I want to take you to the "meat and potatoes" of salt. (pun intended).

Your body has chemicals in it which are **critical** to the functions of life itself. Electrolytes are minerals with electrical charges- they are very small but very powerful.

Major Minerals-Electrolytes

Sodium

Chloride

Potassium

Magnesium

Calcium

These and others are critical for body functions.

SALT

- Needed to generate energy
- Helps cell walls remain stable, cells have walls. Very necessary.

These electrolytes generate and regulate the electricity of your body. Your body operates on chemical reactions and electrical activity. They regulate the muscle contractions-and keeping a proper amount in the body will help avoid muscle cramps and heart rhythm irregularities and abnormal contractions.

Ok, you look puzzled. Think of the **EKG**-*ELECTRO* Cardiogram.

An EKG, also called an ECG or electrocardiogram, is a recording of the heart's electrical activity. EKGs captures a history of the electrical impulse of the heart. The electrical pulse starts at the heart's "spark plug", the sinoatrial (SA) node, and following through the heart middle to the base of the heart muscle. THEN- the heart contracts! One beat!

If there has been any heart damage scare tissue is created, this will interrupt the path of the electrical impulse- and will show up on the EKG. Because electrical impulses cannot easily pass through damaged or dead tissues.

Another reason recorded electrical impulses on an EKG can be abnormal is an imbalance in your critical electrolytes.

A severe imbalance can be life threatening. A severe imbalance can cause the heart to beat irregularly or stop.

So, what I want you to consider; that there are no "*BAD*" electrolytes. Salt -sodium chloride is not *BAD*. It is necessary for life to continue. But too much salt…….ug- not good.

Water follows salt. Your body is always seeking a balance in the water it holds. Water & salt relationship. A little too much salt and healthy kidneys will excrete it. But if this doesn't happen the body will swell up with excessive water. This taxes the entire body-especially the heart. The heart is working hard -pumping blood through wet spongy tissues. Our hands, ankles and legs can swell.

At its worst -we may cough and get short of breath when the excessive water comes into lung tissues. **This is a medical emergency -call 911.**

These are cardinal signs that there's a problem.

This problem needs medical attention.

Consuming the proper healthy amount of salt in your diet will be reflected in your body. No swelling Your physician may prescribe a specific amount of salt you may safely consume daily. And ,if needed, drugs that encourage the kidneys to eliminate excess water. These are called diuretics.

Where is the salt?

Now we will continue this discussion in the next chapter.

2 Hidden Salt

Thanks for following me. Let's be a nutrition detective and find out where the salt is coming from. A reasonable average salt intake -with no health concerns would be about 2300 mg daily. That is not a low salt diet- it's a regular diet. Let's continue. Table salt is under 40% sodium by weight, so a 6-gram serving (1 teaspoon) contains about 2,300 mg of sodium. At first that seems like a lot -but when you are cooking it seems very small, and the taste of salt just "disappears "into sauces and dishes. So, we put more salt in- then—when it gets served- it seems to lack—salt. So, we salt it, again.

A true story. I was in nursing school in a hospital. My patient was lovely lady who was suffering from out-of-control hypertension. High blood.

This dear lady was at her wits end. She reported that she was not using any salt at home at all. She was avoiding chips and bacon, and pickles and olives. She was in tears. And at risk for a stroke or a heart attack. The good doctors had tried everything. As a last resort -she was admitted to a medical unit. That's where were we met.

A few days into her hospitalization her blood pressure stabilized. She was very puzzled.

We talked about this mystery. I asked her to recall a normal day's menu -when she was at home.

She was quick to tell me all the things she didn't have in her house. No pretzels, no beef jerky, etc. OK. , but what was she eating?

So, we started:

Breakfast:

> Eggs, toast -butter with no salt, Polish Sausage.

Lunch:

> Salad, no salt dressing. Bread with no salt butter. Polish Sausage.

Dinner

> Greens, potato with no salt butter, and POLISH SAUSAGE.

What do you see?

Nutrition Facts for Polish Sausage

(Depending on brand and type, of course)

Serving Size: 1 sausage (10" long x 1-1/4" diameter)

Amount Per Serving (approximately):

Calories from Fat: 600

Calories: 800

Sodium: 2000mg

(All numbers are approximated.)

This dear lady had no idea that her very favorite food was the cause of her difficulty.

Ok, you may not like Polish Sausage, but I hope you get the point. You may be consuming more salt than you wish to-unknowingly.

So, let's look at where salt hides. You may not have thought of some of these things.

On a diet? Everyone wants to give you celery.

- One cup -celery 80 Mg of Sodium

That's a lot of salt if your doctor has you on a 1500 mg a day restriction. Is that where you want to use up your salt allowance?

Here's some foods to take a second look at the sodium content- get ready to be surprised!

Vegetables- frozen with sauces

Canned vegetable juice

- Pickles
- Baked goods
- Sauces
- Cottage cheese
- Chinese food
- Canned soups, canned meats and fish
- Some ice creams
- Dairy products
- Shellfish
- Red meat
- All processed meat -unless salt free
- Canned vegetables
- Baking mixes
- Flavorings- marinades
- Butterscotch, caramel
- Salad dressings
- Sports drinks -look at the contents
- Biscuit mixes
- Cake mixes

Commercial bagged snacks, prepared savory, and sweet salty snacks can have high salt contents as well. Nuts,(while they are a wonderful source of nutrition), you need to choose the unsalted variety.

When you see the following ingredients listed on the product nutritional panel you are seeing salt.

These ingredients include salt or sodium-containing compounds, such as:

Soy sauce

Teriyaki sauce

Sodium alginate

Sodium citrate

Sodium nitrite

Monosodium glutamate (MSG)

Baking soda (also called sodium bicarbonate)

Baking powder

Disodium phosphate

I encourage you to take a good look at the nutritional panel on all packaged foods. Good. You need to read the ingredients, also.

Home water softener systems usually use salt to soften the water. This is just lovely for washing clothes -but if you are drinking this water, you are consuming salt.

Many name brand foods, restaurants, and "fast food" chains also post their nutritional facts and ingredients online. You may want to investigate foods you eat often- and save the information for your later reference. There may be a phone app.

Yes, *prepared foods*. That's where we'll go next.

3 Prepared foods

Don't think I want to take away your food fun. Hey! I love eating as much as you do. (Maybe more).

We see many enticing and confusing labels on packaged foods today. Here's what they mean:

Sodium-free or salt-free. Contains less than 5 mg of sodium per serving.

Very low sodium Contains 35 mg of sodium or less per serving.

Low salt, low sodium. Contains 140 mg of sodium or less per serving.

Reduced sodium- this is ambiguous. How much salt was there in the original product?

No added salt- **but may contain naturally salty ingredients.**

Lite salt- the product has been reduced 50% from the original. Ambiguous- how much salt was in the original?

No added MSG- seen at Chinese restaurants. Many of their ingredients may come with MSG within them, however.

According to the United States Food and Drug Administration (USFDA), "Despite what many people think, use of the saltshaker is not the main cause of too much sodium in your diet. **In fact, about 75% of dietary sodium comes from eating packaged and restaurant foods, whereas only a small portion (11%) comes from salt added to food when cooking or eating**" (2016).

Let's look at some common fast foods:

Cheeseburgers

Milkshakes

Fried chicken

Review the nutritional values and you will be surprised at the high content of salt.

Some foods we eat throughout the day contain significant amounts of salt- like baked goods. Breads, cookies, cakes have salt that we may not be able to taste. That's why it's beneficial to read the labels and add up your intake numbers throughout the day (and evening, too). It's normal to be alerted when foods taste salty- like pretzels and chips. But we can be less observant of the salt content of foods which taste sweet or taste as if they need salt.

A four-ounce serving of ground beef has 75 mg of sodium. To my taste it needs salt. So, I cannot judge by taste the content of the salt within my food. Most people cannot. There *are* ways to work on the salt content. That's what we will look at in our next chapter.

Work with me now--- on thanksgiving we see commercially name brand turkey that have a small note on the package. "Contains up to 15% of salt solution". Meats injected with salt water are heavier, and saltier.

When I was a kid, back when the dinosaurs roamed Philadelphia. a question at Thanksgiving was "is the turkey moist?" Today the turkey has no choice but to be moist. The same chemicals that make that formerly low salt turkey now moist and savory are the chemicals you don't need in your precious body.

4 Food Fixes

There are many things you can easily do to reduce the amount of salt you consume. In the last chapter, we looked at food labels and places of hidden salt.

We will now look at what you can easily and cheaply do in your kitchen to impact on your salt intake.

First a story. I was a new nurse working with patients who received commodities. I was told at a university conference that nurses should advise clients who receive commodity canned meats to pour out the broth. Then place the meat in a colander and rinse the meat thoroughly to remove the excessive salt. This sounded great.

In interacting with the patients some listened politely. Others just laughed and laughed. The push back was "that's the best part! We love that broth! There's no way we will rinse off that meat!". Many of these patients were struggling with high blood pressure, kidney disease, and diabetes. They knew what they liked and were not willing to change.

Possibly, the hardest thing in trying to improve our health is changing the way we eat. Our eating habits are rooted in our culture and childhood. There are emotions attached to the food we choose.

Many of us have had to change how we eat. I am not immune from food changes. It's tough!

One way I encouraged myself is reminding myself that I want to **choose health**. That no food tastes as good as feeling healthy. Poor English, but you understand what I'm saying. Our health improvement and good health maintenance depends on making good decisions. Each time we go shopping. Each meal. We can have all the information a university can instill, and still make very bad choices in food.

I am hoping you will make good choices. The ball is in your court. Choices you make now will determine the quality of health you enjoy at present and in the future. Choose wisely.

Vegetables, canned-drain, rinse- then soak in water and rinse again.

Meats and fish, canned- drain, then soak in water and rinse again.

Choose fresh vegetables -prepare them simply, steaming or microwave

Buy meats without sauces, without seasonings that have no added sodium.

Rotisserie chickens in the deli are loaded with salt- take the skin off. I've asked the delis if they could prepare some without the loading of the marinades. I never get an answer. Just a blank stare.

 At home try stuffing chicken with sliced oranges, and some herbs before you bake it.

Someone told me that if I want hot fresh French Fries at McDonald's ask for salt free. They will make them just for you- and then you can -should you wish- just put a light sprinkle of salt on.

If you are eating out -please ask for your salad dressing on the side. Better yet- get oil and vinegar - there's no salt there.

Broiled meats can have a lot of flavor – leave the salt off.

If you add salt when you cook- please rethink . Cooking food sucks up salt, while the taste of salt disappears. When it's served you want to put more on? I am suggesting a new habit to acquire. Abstain from salting food as you cook. Then should you want a bit of salt taste -a very light sprinkle of Kosher salt flakes dusted on top- will give you the taste -with a small bit of salt.

For extra flavor try using citrus juices. Slices of orange add an interesting flair to grilled fish -and chicken.

Get creative with herbs. Each month buy a new one. Experiment, have fun with it. Please shun the spice mixes with salt added. I've seen very expensive mixed, blended seasonings. Only to find the major ingredient was salt. This is very expensive salt! Make up your own blends! The little envelopes with the large variety of "mixes" – marinades, sauces, "seasonings" often are chuck full of salt. Once again -very expensive salt, at that!

Beware of salt substitutes – read the labels. Please. Many contain the chemical potassium. This is a serious mineral l and potentially dangerous electrolyte to add to your meal. It could unbalance your electrolytes or trouble your kidneys. Ask your doctor before you add potassium salt substitute to your diet.

Good News!

As you diminish the salt you eat you will notice flavors you were unaware of previously begin to shine through. People who go on low salt eating plan often, at first, feel the food is lacking in taste. Then in a few weeks they enjoy their food as they never did before- the flavors they never noticed are coming through now. Your usual food can have newly discovered tastes, and pleasures.

5 Blood Pressure

Your body has a closed system of blood vessels. Very much like hoses. Too much pressure and, you guessed it. The hose will bulge and may burst.

Your car has a closed system -the engine -where the oil is. My little car takes 3.7 quarts of oil. Time and time again,-when I get my oil changed it is over filled. This is a type of high blood pressure in a car. If this is not remedied quickly the pressure will blow out the gaskets- (I've had it happen) and the engine in my little car can be seriously damaged. Perhaps destroyed.

If your body's blood pressure is elevated, your entire body -eyes- to toes are being put under pressure. The "gasket" you blow may be a stroke, a heart attack, eye, and kidney damage. Because high blood pressure effects your entire body 400,000 deaths annually are attributed to high blood pressure in the US.

About 75 million people or 1 of 3 U.S. adult have high blood pressure. ONLY about half (54%) of these people have their high blood pressure under control. High blood pressure increases the risk for heart disease and stroke. And other serious and potentially destructive diseases ; kidney failure, eye problems. The entire body is adversely affected. It is the leading causes of death in the US. High blood pressure -the "silent killer" often has no symptoms. The FIRST symptom may be a stroke or kidney damage, as examples. All the negative possibilities are a very long list.

Bottom line: Some things you can have an impact on -some not.

Your genes -ethnicity -certain people groups have more high blood pressure problems.

Age- the older we are the higher the probability- but children with certain kidney diseases can also have high blood pressure.

Weight- The more fat we carry the harder our hearts must work

Diet- see previous chapters.

Stress- we all have stress -we can learn to manage the emotional impact on ourselves. (My stroke was due to over work and long hours).

Lack of physical activity- I recently heard that a death sentence was "a recliner and a remote control." People who have dogs they must walk usually have better health and lower blood pressure. Some studies show that petting a dog can lower some people's blood pressure. (I have a personal story about dogs and blood pressure. Once upon a time I decided to raise Yorkies. Of course, foolishly,

I jumped in at the deep end and got six breeding Yorkie dogs. I don't know about your yorkie -but theses never stopped barking. After enduring one litter- and placing them in good homes—all the dogs -adults and pups. That would be ALL of them. I got a black lab. My blood pressure dropped twenty points. So, I guess it's what you prefer. (Or can endure.)

No surprises here, tobacco use and alcohol can be detrimental to your blood pressure.

Consuming excessive salt can cause your body tissues to swell. ***Water in your body follows salt***

In an attempt to stabilize salt concentrations, and excess water, within your body hands feet, legs and face may swell. The lungs may absorb excessive water and shortness of breath and coughing can occur. This is a medical emergency -call 911.

BALANCE. Healthy kidneys work at dispatching the excess water and salt. Meanwhile the water engorgement of the body's tissues causes the heart to work harder. Blood pressure goes up. It's almost like drowning on the inside.

Something we an impact on- some we live with.

A healthy lifestyle, appropriate sleep, management of personal stress with the company of supportive healthy friends can promote good health.

Look at your diet as an opportunity to impact your health. You can choose to make the changes needed. Ask your doctor what your salt intake should be daily.

Weigh yourself weekly -unless the doctor says more often is better. A shift in more than two pounds is usually water. Call the doctor. Ask- how much weight gain is a concern? And "how much salt am I to have daily?"

Get busy read the labels in your kitchen. Count your salt/ sodium intake and balance it like you do your checking account.

6 Salty Facts

The word "salary" comes from the Latin word for salt because the Roman Legions were sometimes paid in salt. The "salt money," _salarium argentum_, from which get our English word, "salary". Today we may hear someone ask, "is he worth his salt?"

In ancient China salt was very expensive. Chinese royalty was said to commit ritual suicide by eating excessive salt. Early Greeks worshipped salt, no less than they worshipped the sun.

There are at least 32 references to salt in the Bible. The story of Lot's wife who was turned into a pillar of salt when she disobeyed and looked back at the wicked city of Sodom. Pillar of salt, now a common expression. God's encouragement not to look back.

Thousands of years ago the Egyptians used salt to preserve the mummies. Some of which are still intact in museums.

A historic major event was a non-violent protest of the people of India. The tipping matter was the unjust British salt tax, and the British rule in 1930.

The independence movement of India was begun by Mohandas Gandhi. (1869-1948).

The "salt satyagraha" ("march to the sea"). This was a major act of peaceful civil disobedience. Thousands of Indians followed Gandhi 240 miles from Ahmedabad to the Arabian sea.

Turtles. Sea Turtles have unique glands next to their eyes which discharge excess salt.

Do not drink sea water. Seawater contains about 35,000 parts per million of salt-that is a lot of salt. Human kidneys need fresh water to eliminate excessive salt. Consuming sea water can lead to deadly dehydration.

Ok now that you are salivating……

Saltwater Taffy. In Atlantic City, New Jersey there is a walkway. The Boardwalk, with merchants and restaurants and today casinos.

When I was a kid, I could not wait to get to Atlantic City to eat the Saltwater Taffy. The legend goes that an assistant in a boardwalk candy shop substituted sea water for fresh water. Perhaps it's an urban myth- but there are those of us that just love the taffy. It's soft, and sweet, with a hint of salt, almost unperceivable. But it's not sea water. It is table salt, one of the ingredients.

On another note; Blood, sweat and tears all contain salt. There is a serious genetic condition, cystic fibrosis, that can cause salty skin. (And many other problems).

In the Bible, the Lord says in the book of Matthew 5:13 "Ye are the salt of the earth: but if the salt has lost his savor, wherewith shall it be salted? it is thenceforth good for nothing, but to be cast out, and to be trodden under foot of men." He tells Christians are to be as salt to the world.

Salt preserves meats and hinders decomposition. Salt gives taste .Salt is unlike any other chemical. Salt is necessary for life.

Bottom line; you have learned that salt is a critical element for life. You have learned where it is -and if you wish, how to limit it in your diet. You've learned some historic facts. And now you a better able to manage your salt.

Choose good health!

References and Resources

Weinberger, Dr. Myron, Director of the Hypertension Research Center at the Indiana University School of Medicine in Indianapolis and the study's principal investigator. Feb.2001

National Institutes of Health Department of Health and Human Services USA.gov

Study Shows New Link Between Salt Sensitivity and Risk of Death ,2001

https://www.nhlbi.nih.gov/news/press-releases/2001/study-shows-new-link-between-salt-sensitivity-and-risk-of-death

Indiana University School of Medicine's Public and Media Relations Office at (317) 274-7722

NHLBI scientist, contact the NHLBI Communications Office at (301) 496-4236.

*

Katherine Zeratsky, R.D., L.D., Jennifer K. Nelson, R.D., L.D October 16, 2014 Nutrition-wise blog

Like adults, children are eating too much salt, Healthy Lifestyle

Nutrition and healthy eating, Mayo Clinic

http://www.mayoclinic.org/healthy-lifestyle/nutrition-and-healthy-eating/expert-blog/children-eating-too-much-salt/bgp-20113918

Minnesota — Rochester

Mayo Clinic

200 First St. SW

Rochester, MN 55905

General number

507-284-2511

*

Mayo Clinic Staff, April 2016 Sodium: How to tame your salt habit, Healthy Lifestyle

Nutrition and healthy eating, Sodium: How to tame your salt habit, Mayo Clinic

http://www.mayoclinic.org/healthy-lifestyle/nutrition-and-healthy-eating/in-depth/sodium/art-20045479

Minnesota — Rochester

Mayo Clinic

200 First St. SW

Rochester, MN 55905

General number

507-284-2511

*

Wikipedia contributors. Salt in Chinese history. Wikipedia, The Free Encyclopedia. September 11, 2017, 08:50 UTC. Available at: https://en.wikipedia.org/w/index.php?title=Salt_in_Chinese_history&oldid=800057228.

*

Wikipedia contributors. Salt. Wikipedia, The Free Encyclopedia. September 18, 2017, 07:17 UTC. Available at: https://en.wikipedia.org/w/index.php?title=Salt&oldid=801195158.
https://en.wikipedia.org/wiki/Salt

*

Wikipedia contributors. Salt water taffy. Wikipedia, The Free Encyclopedia. September 19, 2017, 13:53 UTC. Available at: https://en.wikipedia.org/w/index.php?title=Salt_water_taffy&oldid=801407669.

https://en.wikipedia.org/wiki/Salt_water_taffy

Morton Salt

http://www.mortonsalt.com/salt-facts/

Consumer Inquiries

[In-Home Use Products]

T: 800.725.8847

E: saltinfo@mortonsalt.com

*

Moldon Crystal Salt Co.

http://www.maldonsalt.co.uk/About-Salt-Salt-interesting-facts.html

http://www.maldonsalt.co.uk/About-Salt-Salt-an-amazing-history.html

Maldon Crystal Salt Company Limited

Wycke Hill Business Park, Maldon, Essex, England, CM9 6UZ

Telephone: +44 (0)1621 853 315   Facsimile: +44 (0)1621 858 191

Email: info@maldonsalt.co.uk

*

Perlman, Howard, Dec 2016 , Why can't people drink sea water?

The USGS Water Science School

U.S. Department of the Interior | U.S. Geological Survey

http://water.usgs.gov/edu/qa-seawater.html

USGS FAQs Homepage

Call 1-888-ASK-USGS (1-888-275-8747)

ABOUT THE AUTHOR

Suzanne C. Smith R.N., born in Philadelphia, Pa. has enjoyed serving others with nursing care and strong patient advocacy. She came up through the ranks- first as a nursing assistant- then an LPN, then a Registered Nurse. She earned a CPHQ- Certified Professional in Healthcare Quality in 2004. A graduate of the University of the State of New York- Regents College, Albany, NY. and Pima Community College, Tucson, Arizona, additional studies; South Dakota State University-Nursing and University of Pittsburgh.

Smith has been honored by Congressmen and other community leaders for work with senior citizens. Smith has worked in the medical departments of two HMOs in the role of Utilization Review, Prior Authorization, Concurrent Review, Quality Assurance, and Case Manager. She also worked in those roles within a private hospital and Home Health. She has been a board member of the United Way- Healthcare Allocations committee, served in the Red Cross, and numerous other volunteer positions.

www.ingramcontent.com/pod-product-compliance
Lightning Source LLC
Chambersburg PA
CBHW050027230526
45470CB00003B/1162